THE DESERT BIOME

Colin Grady

Enslow Publishing
101 W. 23rd Street
Suite 240
New York, NY 10011
USA
enslow.com

WORDS TO KNOW

climate—The weather conditions in a place over a period of years.

coastal—Having to do with land that is next to a sea or an ocean.

community—A group of living things that share the same area.

continent—One of the seven great masses of land on the earth.

dunes—Mounds of sand that are formed by the wind.

evaporate—To change from a liquid into a gas.

mammals—Warm-blooded animals with backbones and usually with hair.

reptiles—Cold-blooded animals with backbones and usually covered with scales.

semiarid—Having only a little rainfall.

CONTENTS

The desert biome covers about one fifth of the world.

The Desert Biome

A biome is a community of plants and animals that live together in a certain place and climate.

There are different types of biomes, such as forests and oceans. Deserts are another kind of biome.

Deserts are found all around the world. The desert biome covers about 20 percent of the world's land surface. Most deserts are covered with sand. There are four main types of deserts: hot and dry, semiarid, coastal, and cold.

Hot and Dry Deserts

Hot and dry deserts are found in Africa, southern Asia, North and South America, and Australia. Temperatures in these deserts can reach 120 degrees Fahrenheit (49 degrees Celsius). These desert areas are so hot that sometimes rain evaporates before reaching the ground.

The areas in tan are the world's deserts.

Cold Deserts

Cold deserts are found in parts of Asia, North and South America, and Antarctica. The summers are short and have light rainfall. Summer temperatures are around 70 degrees Fahrenheit (21 degrees Celsius). Winters are long and cold with temperatures as low as 28 degrees Fahrenheit (-2 degrees Celsius). Snow falls throughout winter and sometimes during summer.

An Oasis in the Desert

An oasis is an area of land in a desert where water comes to the surface from deep underground. Plants and animals can live in an oasis because of the water there.

Semiarid Deserts

Semiarid deserts are mostly dry. However, there is light rainfall in the winter. The days are hot and the nights are cool in a semiarid desert. The cooling of the night air makes dew, small drops of water from cool night air. The dew covers the ground, the animals, and the plants to provide water for all of the living things.

Semiarid deserts can be found in North America, Europe, and northern Asia.

The Driest Desert

The Atacama coastal desert of South America is the driest desert on Earth. Dry periods have lasted up to forty years.

Red Rock Canyon is a semiarid desert in Nevada.

Coastal Deserts

Coastal deserts are mostly found on the western coasts of continents, such as Africa and North and South America. Dry, cold ocean air blows over the coastal desert land. These winds often form large dunes, loose piles of sand that can be hundreds of feet tall.

Summers in coastal deserts are long and warm. Temperatures go from about 55 degrees Fahrenheit to 75 degrees Fahrenheit (13 degrees Celsius to 24 degrees Celsius). Chilly fogs come in from the ocean during winter.

The Namib Desert in Namibia, Africa, is a coastal desert.

The mesquite tree survives in the desert because its roots reach far underground for water.

How Plants Survive in the Desert

There is not much water in the desert, but some plants are still able to live there. These plants have special ways of getting and storing water.

The cactus stores water in its roots, leaves, and stems. Other plants, such as the mesquite tree, have long roots that get water from deep beneath the ground.

Prickly Plants

Plants also have other special ways of living in the desert. Some, like the cactus, grow spines or hairs. These spines and hairs shade the plants from the hot sun and protect them from being eaten by animals.

The spines of the cactus protect it from the sun, as well as animals.

Animals at Home in the Desert

The desert is home to many types of insects, spiders, reptiles, birds, and mammals. Most of these animals rest during the day and hunt at night. They often make their homes in holes in the ground where it is dark and cool. After the sun goes down, they come out to look for food.

Cape ground squirrel

Living in the Desert

Animals that live in the desert usually have something special about them that helps them survive in this dry biome. A fennec fox has very large ears that help it stay cool. The cactus wren is a bird that builds its nest right in the cactus. It gets almost all of its water from the food it eats.

Gila monsters have a strong bite, which can be deadly to the animals they hunt for food. They are found in the Mojave, Sonoran, and Chihuahuan Deserts in North America.

The Ships of the Desert

Camels can go seventeen days without water. They are able to drink lots of water at once and then store it for a very long time. Their wide, soft feet allow them to walk easily on desert sand.

How People Live in the Desert

People have found ways to live in the desert. Some desert areas are near water. People can live there and grow crops like corn and beans. They can also raise animals when deserts have grassy areas where the animals can eat.

The Growing Desert

The world's deserts are getting bigger. Deserts spread when animals eat the plant life, grassy areas dry up, or the soil is

A man stands in his desert home made from adobe.

The desert biome is home to a unique group of animals and plants.

Building Homes in the Desert

People have used adobe for building houses in deserts for thousands of years. Adobe is made from clay that is baked in the sun. An adobe house is warm during cold nights and cool during hot days.

hurt by too much farming or mining. Each year, more than two million acres of livable land are lost because of spreading deserts. The Sahara in Africa is the largest desert in the world. In the past fifty years, the Sahara has spread to the south, covering more than twenty-six million acres!

We must take care of the land and make sure that plants, wildlife, and people can all continue to live and grow.

ACTIVITY
DRAWING THE DESERT

You have learned about some of the different plants and animals that live in the desert biome. Now it's time to create your own.

1. On a large blank piece of paper, draw a desert scene. Begin with the land. Deserts are mainly sand, depending on the type. There may be dunes and underground burrows, as well. How about a water source?

2. Now add plant life. You can draw your plants or find images online and print them out. Show at least two different kinds of plants that you learned about in this book, or you can research some online. Label each one.

3. Add your animals—at least two mammals and two reptiles. Again, you may choose to draw or print out your pictures. Label these, as well.

4. You have completed your own desert biome!

LEARN MORE

Books

Duke, Shirley. *Seasons of the Desert Biome*. Vero Beach, FL: Rourke, 2014.

Johansson, Philip. *The Desert: Discover This Dry Biome*. New York: Enslow, 2015.

Patkau, Karen. *Who Needs a Desert?* Toronto: Tundra Books, 2014.

Profiri, Charline. *Guess Who's in the Desert?* Tucson, AZ: Rio Chico, 2013.

Websites

Ducksters

ducksters.com/science/ecosystems/desert_biome.php

Interesting facts about the desert biome.

Kids Do Ecology

kids.nceas.ucsb.edu/biomes/desert.html

Photos, facts, and links to information about the desert.

INDEX

Published in 2017 by Enslow Publishing, LLC.
101 W. 23rd Street, Suite 240, New York, NY 10011

Library of Congress Cataloging-in-Publication Data
Names: Grady, Colin.
Title: The desert biome / Colin Grady.
Description: New York, NY : Enslow Publishing, 2017. | "2017 | Series: Zoom in on biomes | Audience: Ages 7+ | Audience: Grades K to 3. | Includes bibliographical references and index.
Identifiers: LCCN 2015048569| ISBN 9780766077720 (library bound) | ISBN 9780766077584 (pbk.) | ISBN 9780766077614 (6-pack)
Subjects: LCSH: Desert ecology--Juvenile literature. | Desert animals--Juvenile literature. | Desert plants--Juvenile literature.
Classification: LCC QH541.5.D4 G7175 2017 | DDC 577.54--dc23
LC record available at http://lccn.loc.gov/2015048569

Printed in Malaysia

To Our Readers: We have done our best to make sure all website addresses in this book were active and appropriate when we went to press. However, the author and the publisher have no control over and assume no liability for the material available on those websites or on any websites they may link to. Any comments or suggestions can be sent by e-mail to customerservice@enslow.com.a

Photo Credits: Throughout book, Alex Belomlinsky/Getty Images (lizard icon, cactus icon, camel icon), Natural_Warp/Getty Images (desert background); cover, kavram/Shutterstock.com; p. 4 Marques/Shutterstock.com; p. 6 Joingate/Shutterstock.com; p. 9 Elena Khanyukova/Shutterstock.com; p. 11 Ev Thomas/Shutterstock.com; p. 12 iStock.com/Stockcam; p. 14 Anibal Trejo/Shutterstock.com; p. 15 Grobler du Preez/Shutterstock.com; p. 16 Kris Wiktor/Shutterstock.com; p. 17 Banana Republic Images/Shutterstock.com; p. 19 Luis Davilla/Getty Images; p. 20 Villiers Steyn/Shutterstock.com; p. 23 Dorling Kindersley/Thinkstock.